# Gravity Explained

Edward Johnson

# It's cause
## Vs
# A Mathematical Model

For

Thomas Edward Johnson and Fabiola

All rights reserved

Edward William Johnson

30th January 2013

**ISBN-13:
978-1482327243**

**ISBN-10:
1482327244**

With thanks to Lt Cdr James A Mitchell RN & Gareth L Dean PhD

Edward Johnson

# Gravity as an Operating Mechanism

[ Newton & Johnson ]

**T**heory **O**f **N**ew **E**merging **S**pace

Based on the belief of Sir Isaac Newton that Absolute Space has validity when identifying the cause of

**UNIVERSAL GRAVITATION**

# 1 Introduction

Gravity as an identifiable force or mechanism is one of the greatest physical mysteries alluded physicists since Euclid. This paper reintroduces and reconfirms Sir Isaac Newton's idea and belief that Space provides us with a 'Background Absolute', later discounted by Albert Einstein. It is explained here by the constant emergence of New Space. This same mechanism not only describes how gravitational information is exchanged but also to, the ability of a photon to transmit and transit in space and determines the well known velocity 'C'. This is central and a key feature in this paper. Both Gravity & Light are subject to this fundamental space mechanism which determines the abilities of both. It introduces the concept of 'Primary Space' with a constantly emergent population of 1 dimensional time points which is the physical key to unlocking this long lasting mystery. If this philosophy is correct then it is necessary to change our current audit of spatial dimensions, and remove temporal time as being one of them. The details presented provide explanation of this system and dependent upon the following layout of dimensions. [Ut, x, y, z] and temporal time is excluded from these 4. [ Ut is the symbol for Universal Time Constant ] this being the Primary dimension providing Newton's Absolute.

The mechanism described here is closely associated with Time, holomorphism, information and entropic causes. Erik Verlinde in his 2010/11 paper describes gravity on the basis that space is emergent and causes the exchange of information, but does not actually explain the mechanism of this emergence and is dependant upon strings to transport the information. The topic of emergence is covered in this paper and is divergent from his particular notions and presents an alternative view..

The subject of the concept being proposed may be considered to be a unique approach to unlock the mysterious cause of gravity. With the exclusion of open strings, closed strings or particles. It is totally dependent on the ability of

Absolute Space to cause an ongoing and constant inflation of its manifold consisting of and production of 1D points of time. It also describes that the universe is dependant upon only 4 spatial dimensions for the entire mechanism to work, and indeed all of nature.

## 2 Primary Space

Primary space may be considered as a 'Constantly Emergent Manifold' of 1 dimension not only to provide a mechanism for gravitational information exchange, but also may even go further and provide a candidate for a complete Unification Theory.

In reading and discovering the new ideas in this book we have to adjust our belief that we exist only in the 3 Euclidian spatial dimensions of spacetime. For a long time the idea of the 4$^{th}$ dimension was popularized in cinema. It is currently thought to be a dimensional domain for temporal time. Primary Space should occupy this title not temporal time. So we realise that we exist in the 3 spatial dimensions of spacetime but also the Primary manifold which occurs as a background absolute which is constantly emerging in all directions being the 1$^{st}$ dimension.

Primary Space may be described as a universal manifold having a unity dimension which has a construct of isometric Time = [ 0 ]. Then reordering our current understanding of 4 dimensions **[ w, x, y, z ]** where [ w ] is currently thought to be temporal time. To define the cause of gravity it is necessary to reconsider and change this historic expression to become [ 1 of Absolute Primary Space of Constant Time and 3 of spatial dimensions ] and temporal time having only a application as confined by 3D of [ x, y, z ].

Absolute Primary Space is expressed as [ Ut ] meaning Absolute Universal Time. So the new dimensional expression becomes **[Ut, x, y, z ].** or shortened to [ Ut,3D ]. It does not change the application of Euclidian dimensions it only displaces temporal time and ceases to be known as a separate dimension.

Temporal time is non constant and location dependant measurement of local time which can only occur providing matter exists first providing the ability of 3D dimensional distances, and varies according to velocity and gravity affects on it. One cannot have temporal time in a field where no 3D references exist so I have left this out of the above simple expression as having no application. However, we need to use it in the description which follows to aid our understanding and visualization of the mechanism of gravity as presented herein.

## 3 Universal Time [ Ut ]

Universal time has a construct manifold of time constant space. This manifold may be thought of as a static 1D field. However, is it not because it is being constantly populated by zero points of ( 0 or 1 ) entities of point time. These points form a constantly growing field rather like an expansive foam. This manifold is being added to in every cubic volume of our cognition of 3D spacetime. However, one has to be careful not to think of the manifold as 3 dimensions as it is not. It has an isometric 1 dimension, and is not space in the normal spacetime sense, but has only a void of Absolute constant time. This is where we have to separate the following two identities:

1. [Ut Primary Space TC ] & 2. [ Spacetime x,y,z TV ]
   i.e. TC is time constant and TV is time variable

Universal time [ Ut ] exists as the manifold of Primary Space and is totally separate entity from 3D spacetime thinking. Primary Space can exist in absence of spacetime, and spacetime cannot exchange gravity information in absence of it. The Primary Space may be thought of an artist's palette onto which he manifests the paint and mixes the colour. They are completely separate entities. He does not place the palette onto the canvas.

Despite the fact that they are separate there is interaction between them. [ Ut ] causing matter to occur according to its background rate of population. Where it's kinetic energy to be converted and formation of matter. Also, [Ut] continues to be populated once matter has been formed. This constantly emerging New Space is being created second for second in temporal time and is responsible for the conveyance and transit of information around and through our understanding of 3D spacetime.

The sketch below illustrates the spacetime curvature of spacetime due to existence of mass in it due to GTR. This is an historic understanding and we now visualize it in a different way. But for illustration purposes I use it here.

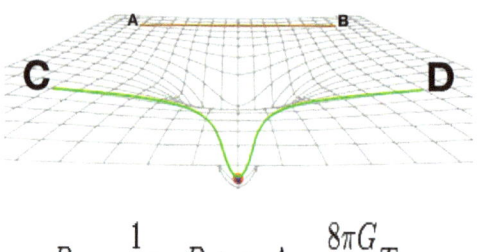

$$R_{\mu\nu} - \frac{1}{2}g_{\mu\nu}R + g_{\mu\nu}\Lambda = \frac{8\pi G}{c^4}T_{\mu\nu}$$

In this image our understanding of spacetime is curved in the above sense where matter is present due to the theory of STR. This illustrates the [ virtual effect of gravity ] by the deformation of spacetime. As defined by Einstein's field equation and Newton's classical theory of gravity. This vision is not really correct although it holds good for teachers to explain the effect of gravity around a mass, without giving reason what actually causes the gravity in the first place. Now we define gravity of a given body by its proportionality to its energy and quantum of matter.

The resultant gravity outward/inward to a body is subject to the inverse square law. i.e. the further you are away from the body the weaker the gravitational affect. So when standing close to that body one may think that a pencil point is acting on your skin and not the blunt end. The gravity is the same but divided by the area of the blunt end, so it feels less the further away from its source. This phenomena is currently well understood.

## Proposed limitation of Einstein's notion of Special and General relativity.

Einstein's symbolic equation below is his General Theory of Relativity and determines how 3D spacetime will curve and then in his STR how curved spacetime tells matter how to move. This does not however explain in any way what Gravity is or caused. The theory is totally relativistic in 3D + temporal time, and may be thought of as a tool to only aid the 3D understanding how his vision of spacetime and 3D matter in it.

On the right hand side of the equation [ $T\mu\nu$ ] is what is known as the [Stress – Tensor $x^0 = t, x^1 = x, x^2 = y$, and $x^3 = z$, where $t$ is time in seconds, and $x, y,$ and $z$ are distances in meters]. It is totally relevant and dependent upon Euclidean vectors and temporal clock time, which should not be considered as a true vector in this book and one which is only subject to the ability of matter and energy. E.g. 3D distances and displacement between the interacting energy and matter.

$$G\mu v = 8\pi T\mu v$$

If we conclude we live in a universe with 2 entities of time acting as totally separate vectors. Then if we consider the symbolic equation below acting in true 4D it eliminates consideration of spacetime curvature completely. [AR ] in this expression means Absolute Relativity. If you consider in this paper the explanation for the cause of gravity has the potential and cause. We then have no need to complicate things and think of space time bending to explain it. On the basis that we now know what gravity actually is and we can illiminate the historic mathematical model for it. However Sir Isaac Newton law still holds good by simply changing the values of mass in his equation for Time Constant [1' or 0's] of information. The [$r^2$] below retains our 3D understanding of Euclidian space and entropic effects ( weaker at distance ) of the gravity produced. The equation immediately below is Newton's law of gravitation but simply modified to replace mass for information. The lower symbolic equations represents the 4D of the Primary space + Euclidean 3D. This could well be the equation for Everything. [Ut ]provides the mechanism for gravitation and creation of matter in the first place, and 3D, observation temporal time & velocity. One then has a system to create matter and evolve in it.

$$AR = G \frac{01_1 01_2}{r^2}$$

$$AR = Ut3D$$

The application of this theory is acting through a plane of 1D of Primary Space without any need to justify and explain a curved space and time in 3D + temporal time. The force of [G ] in this case is simply acting perpendicular to the body or the particle diameter. Illustrated by the ripple image on page 9 (actually in 2D presented for explanation purposes ) between 2 bodies.

The force acts in a perfectly straight line through the medium of the 1 dimensional plane. The force of the mutual gravity is a sum of its [1 and 0 points]. Then all the relevance of the General Theory of Relativity ceases to have any further case or application. Then the door of scientific understanding is suddenly wide open. The following explanation presents more details to understand this natural phenomena.

# 4 Absolute Primary Space Manifold & Production of Emergent 'New Space'

This is an important chapter as it introduces this new notion and attempts to explain the mechanism of the 1 dimensional manifold which we have ignored since Newton having stated that space has an independent and Absolute construct.

Here is a simple spherical point or we can think of it as a 1D flat solid circle or another geometric, the actual definition of the shape does not really matter. This entity cannot be directly observable in 3D space and most likely existed prior to the so called Big Bang. It is constant time zero and exists in a manifold which has a construct of exactly the same 1D entities as illustrated by the populated block below.

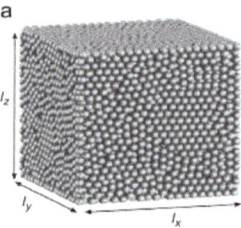

In the above image we have built a cube [virtual space in 3D] of identical points. But the entire structure does not have 1x, 1y or 1z. It is a unity dimension which we call [ Ut ]. As it is composed of points of time which are dimensionless and have no distances, or faces to measure

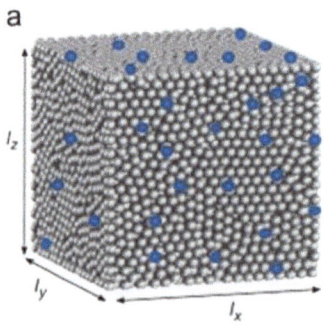

In this image [ Ut ] above has enlarged its volume, constantly emerging with more zero points of time. The normal distances 1x, 1y and 1z have 'virtually' changed. But [a] continues to have no dimension.

Its original virtual volume has only become more populated with zero point time entities. There is no change it its dimension as it has non to change. It is an invisible enlargement to us which is participating and interactive with x, y, z spacetime. This is what Sir Isaac Newton could have had in his mind had he lived long enough.

He referred space as providing a background Absolute. Unfortunately, he did not think that it had a construct of time constant points or indeed was a dynamic, and thought of it as a static medium. This was later changed and investigated by Michelson and Morley, who thought it was a wind. Plus numerous other notable scientists who referred to it as the Aether. Then later in Einstein's career having discounted Newton's notion of an Absolute space then reinvented it calling it the Absolute Aether, as something was missing. Since then we have thought of spacetime as an absolute having only 1 identity. I propose here that this is not the case and Newton's notion was probably correct. And we have to bypass Einstein's notions of spacetime and cause it to be modified to incorporate Newton's concept of an Absolute. And also to refer to the 4 dimensions as [Ut, x, y, z ]+ temporal time as explained earlier.

To conclude in brief at this point, we exist in space with two identities: of Absolute space + spacetime. The primary space has a background construct of constant time, and constantly being populated which causes it to be a dynamic and increase its virtual 3D volume. The rate of its emergence is detected by and the ability of light speed [C]. The mechanism and constantly emergent manifold is what determines the upper velocity of light and indeed anything to move whatsoever. Be that a photon or wind against a petal on a flower.

New Space is being created second for second when measured by the upper value of [C]. Nothing can move in x, y, z spacetime where [Ut] does not yet exist or not emerged. Everything is dependent upon the continuous population of constant time to form New Space. Think of this as being able to walk on a conveyor belt. If the conveyor belt stops moving we lose our ability to walk. The belt has to constantly move and emerge otherwise we cannot.

How to visualize the emergence of New Space is illustrated in the following sketch.

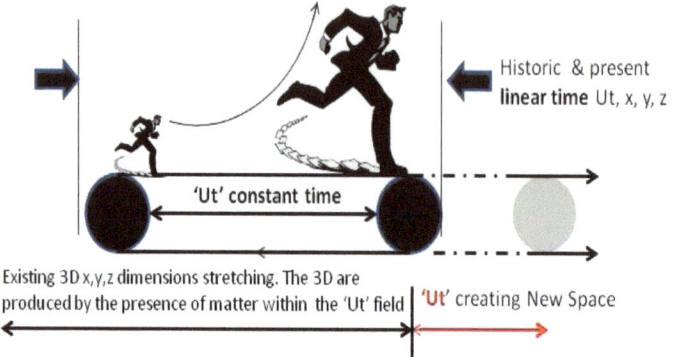

The man can only move if the belt is emerging. The belt in this case is not only moving but growing longer, producing more of the belt. Both have to happen in order for him to have any ability of movement. In the above picture he is able to run only as fast as the belt is moving. He is also at a point of the vertical black line. This line has to be constantly moving away from him, if it does not then although the belt may be moving he cannot move. He has to have constantly supplied New Space. This is what the Primary dimension of Absolute Space [Ut] is providing us.

It is constantly opening enabling more invisible 1D space for us to move second for second. We cannot move in 3D time & space, if [Ut] is not being constantly populated causing its ongoing emergence and ability.

# 5 Gravity

The following is Newton's well know law of gravitation and the Force acting on both bodies is subject to the combined ratio of masses and inverse square law between them.

$$F = G\frac{m_1 m_1}{r^2}$$

## Cause of Gravity by a ratio of exchanged information

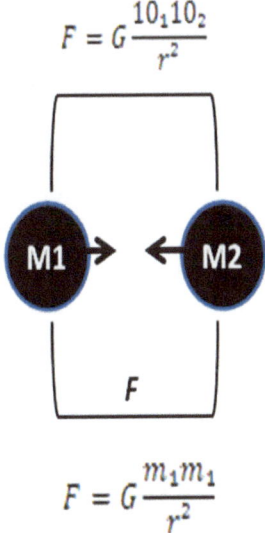

Effect of gravity caused by mass

In the above illustration it identifies the new expression for the [**cause of gravity**] and below the [**effect of gravity**[. The top equation acts through the medium of Newton's Absolute Space, consisting of 1 dimension of emerging 'New Space' and the lower through traditional historic 3D vision of Einstein's curved spacetime + temporal time.

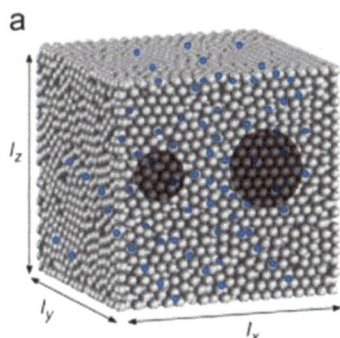

In the above image there are two bodies shown semi transparent and immersed in a constantly populating and emerging volume of 1D Primary Space. These bodies do not have any mass in the macroscopic 3D sense. They are only identifiable by their energy and quantum of matter, which may be part of quark or photon etc. They are moving independent from each other and have vibration. However, they can only move and vibrate if New Space is being provided in keeping with the man on the conveyor above. [ **They cannot move in historic space which emerged in the previous second or infinitesimal part of a second as it has moved away from them].** This is the crucial part in this explanation and comprehension of what causes gravity and determination for the velocity of light as we know it.

One can think of these bodies existing in [Ut] in various forms and scales as large open cell sponges. On the basis that the relative size of a sub atomic particle compared with a 1D zero point of time one could say is infinitely huge. The 1D dimensional points are not aware of them as solid unique entities but infinitesimal components of atoms which have a quantised ( amount ) of energy state, movement and including its vibration and amplitude.

These constant 1D time points interact with their state and behavior. The manifold is increasing in volume in all directions uniformly at the rate of value [c] away from each entity which on the macroscale we refer as mass. This virtual 3D metric volume increase of the 1D absolute manifold is not related to the speed of light – light speed is determined by it. The emergence of New Space is limiting its upper velocity. And a photon in isolation is our detection and measurement of the emergence of New Space population rate being created.

As each new 3D virtual volume of [Ut] is produced the information relating to each 'sponge' ( energy and matter ) interacts with it as it is being created. One could say analogously similar to a magnetic tape. There is a relationship with the state of energy and matter and the production of New Space as it is produced. As this New Space continues to be produced endlessly, it is influenced by the 3D energy and matter objects in it. Which again referring to the magnetic tape collects all the information into its 1D time manifold.

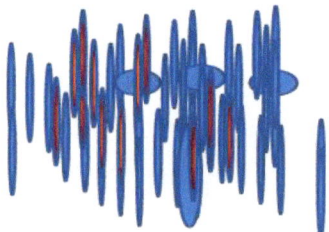 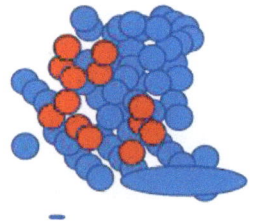

The images above shows the idea of constant time points being modulation represented by the red new dots and slits. There are no particles it is a continuous flow of New Space and has no visible grain. In the same way one can consider a magnetic tape as a smooth, uniform and homogenous surface. As each [Ut] of 1D time point appears they are modified. Remaining as [0] or revalued as [1]. They subsequently and inadvertently record the measure of energy and matter in that particle. This produces a constant stream of updated information which moves constantly outward from a particle, which is switching the [ 0's to 1's ].

This constant stream of information ( the magnetic tape spool ) may be many billions of light years 3D scale in distance as they continue to propagate. If a particle the source of that information varies, moves or ceases to exist, all that information is recorded, just like a singer into a microphone connected to a tape recorder. [Ut] records all information from its beginnings to its end, and is constantly transmitted into the enlargement of Primary Space the 1D Absolute and primary dimension. Every man that has been murdered absolute space knows who the murderer is! It is offering us far more than gravity.

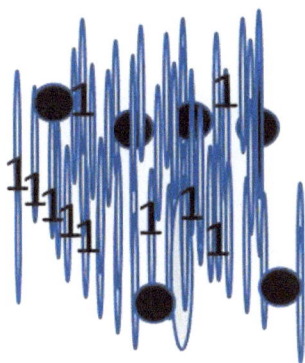

This image illustrates how New Space is emerging and the 1D points of absolute time remain as a [0] or interact and become [1] according to the quantum of energy and matter in its vicinity. This happens in every second of our temporal time hence all the information in and on the Earth is recorded. Every tree, every person, atom, all the Quarks, Hadrons, and particles species in the Standard Model of particles existing in every atom - entire planet and universe for the last 13.7 billion years. This seems like a lot of temporal time but in the absolute space it does not make 1 second! It copies all the information. This is the ability of New Space [Ut ] as it is being constantly produced. This occurs through every single atom in the universe whether that atom is bound to a black hole or an earthworm.

The propagation, population and emergence rate of [Ut] is witnessed by the current value of [C]. Close quarter infinitesimals and bodies billions of light years away all correspond. The information will always be historic with respect to each other, as is gravity and light moving between them. In the case of the infinitesimal sub atomic particles the 3D attraction is dependent upon the electromagnetic affect and independent from Newton's law, modified here. However, in order for two particles to join New Space must be created between them, otherwise they have no ability to move or vibrate.

The mechanism being described here is also the engine for the universal dilution of matter in the entire universe.

When we think of gravity in terms of 'mutual gravitation' it is simply explained that all bodies are producing information about themselves as New Space is being created isometrically everywhere in all directions and between them all, but confined to a 1 dimensional manifold framework. A huge universe which is flat in terms of time. ' **Tempus fugit discus'**, time passes flat in the sense that is only has 1 dimension. But as far as you and I are concerned it is not as we makes sense of the universe in Euclidian 3D.

This image represents the mutual deformation of our understanding of an object or attraction to another by their sum total of their joint masses, meaning their energy and matter quantum. This is how the masses prefer to join and not according to historic representation of a curved spacetime.

They are not deforming because they know of each others state independently by means of a particle transfer such as a graviton, they are deforming on the basis of the information being provided by each of them is through the plane of [Ut] of 1D constant time. If a body has information containing more 1D points of [1] than [zeros] than its neighbouring body then that body will have more gravity. It shall possess the dominant gravity.

2D resultant gravity representation, where information is exchanged in a 1 dimensional plane not via GTR

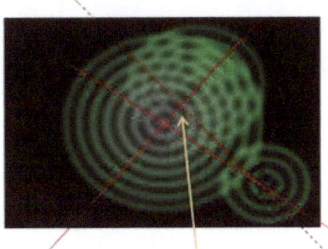

Newton's Absolute Space point acting independently from the affects the mutually rotatingmasses

This is a simple ripple tank image illustrating a well known affect in our experience of 3D of propagated ripples in a liquid and how they interfere. In the case of [Ut] the mutual mixing of information [ 0 and 1 ] initiates the attraction manifesting 'The Gravity'. If the bodies are closely coupled like our Moon and Earth the information density is greater in keeping with the 3D inverse square law. Conversely should they be billions of light years away substantially weaker – however that historic information continues to move outward, analogous to a radar beam. As far as [Ut ] is concerned the information is being transferred through a flat plane. Like the ripple tank above. How we experience and visualize it by the previous image of two 3D bodies being pulled into each other.

The mutual attraction of gravitation is achieved by transfer of the information relating to its energy and matter contained within their structure, and when we revisit Sir Isaac Newton original notion of the apple. This is the mechanism at work. New Space being created constantly between the earth and apple is transferring the information of their respective structure. In the case of the apple it has substantially less [1's] of information than the Earth. The apple by comparison is somewhat substantially dilute of Gravity as measured against the earth and is pulled toward it more than the Earth is pulled to the apple.

When Newton's apple disconnected from the parent tree it was obliged to fall to toward the earth as it is the most local and dominant of all the information relating to energy and matter which is influencing it.

[Ut ] Emergent New Space of 1 dimensional constant time

In the image above shows the earth as a solid object internal to the grid of exterior spheres. The information of [Ut] encompasses the entire solid content of the earth, and collects information on the infinitesimal scale from it. The resultant image is identical to the actual 3D solid size form and propagated into the 3D of spacetime as 1D object. It looks like a sphere but it is built in 1 dimension. The spherical face is not moving outward like a ripple in the tank of water, the information is moving away from the source as New Space is being created. It is not a wave it is an isometric evolution of information in 1 dimension, like the magnetic tape analogy.

In the above image depicts Earth as it was approximately 4 billion years ago when it was completely molten. At that time is had a significantly higher energy state than it does today illustrated by the cooler previous image. In all that time New 1D Space has been constantly emerging and recording all that information and conditions in every temporal second of the evolution on Earth.

The above picture and effective gravity is what is seen by another object some 4 billion light years away from us today. Earth would appear as an object with relative greater gravity due to its higher thermal energy state. That is, if the information density was not subject to Newton's inverse square law of gravitation, and become immensely weakened gravitational intensity.

It is now necessary to update and substitute the symbol for [ $M$ ] mass for [ $E$ ] energy in Newton's original equation, and [ $mQ$ ] for matter. So the distant body which has just received our Earth's primordial information is the inverse of its [ $E+mQ$ ] not the mass according to Sir Isaac Newton's original equation. Matter is a term which relates to the total quantized entity of materials available to be measured in a body, including the infinitesimal mass less versions of matter. Not to be confused with mass of that body which involves a different measurement and identity.

During this explanation it is important to visualize that the gravitational information is in the form of a regular temporal second for second duplication of the Earth's $[E+mQ]$ reference even though it be 4 billion light years reach ,or 8 billion light years of information diameter. Earth may be considered as having a second for second duplication of itself across that huge distance, projected into the 3D spacetime accounting for every second right up to its current date. Not one second of Earths historic information has been lost in all that historic temporal time.

If our belief that Einstein's field equation laws are correct, which they are but only relative to 3D [ x, y, z ] spacetime and Minkowski space of temporal time.

Einstein and Newton's views were and are at variance. The purpose of this book is to re invent and redefine the application of Newton's law of gravitation with reference to Energy and Matter and only relevant to Gravitation not the 3D STR & GTR spacetime mathematical model for it.

Therefore, if one elect's to remain in the School of Einstein one will be forever tethered to 3 dimensions of spacetime including Minkowski's temporal time. However, should one elect to join the Newton Johnson school you will be able to understand a new opportunity to comprehend what actually causes Gravity based upon his Law of Gravitation but substituting his values and substitute mass for energy and matter and translate that into pure relative information. In doing so will open a new door and reconsider the existing audit of dimensions. With the possibility of 4 actual spatial dimensions and the principal one is the Primary dimension which Newton thought of providing the Absolute background of Space. Which was discounted in and around the early part of the 20C due to Einstein referring to it as "only of "hypothetical interest with no application".

This is Newton's Classical 17C law of gravitation according to mass of on object.

$$F = G\frac{m_1 m_2}{r^2}$$

If we substitute $m_1\, m_2$ for:

Energy and quantum of matter $E$ and $mQ$

Newton's 21C Law based on the ability of the emergent 'Absolute Background of Space' and Gravitation:

$$F = G\frac{EmQ^1 EmQ_2}{r^2}$$

[r²] in this equation is constantly increasing in length at the value of [C]. This is a new equation for Gravity and the mechanism for its transfer is the Primary dimension [Ut] which creates New Space second for second. The $E$ and $mQ$

interact and cause the constant [0] time points to remain a [0] or change to [1]. This is the mechanism of gravity. Then these [0] and [1] points interact with a neighbor body information such as Newton's apple and hence determine the ability , ratio and experience of the force of mutual gravitation.

This equation is thus up to date and represents the gravitation experienced between 2 bodies in 1 dimension. Einstein's field equations may continue to be used safely but only relevant to a 3D visualization of the affect of the bodies which are actually probably influencing each other in a plane of 1D.

## 6 Gravity as Newtonian Johnson operating mechanism

For gravity to work it must have a single mechanism whether we understand it or not. Historically and currently this is a complete mystery, and something which happens around us and through us everyday which we simply cannot fathom. Albert Einstein created and ingenious set of mathematical models which provides us with a visualization of its effect. Unfortunately they do not provide any indication what actually causes it. So we are far distant from comprehending what is this operator. Sir Isaac Newton and Einstein are both important participants in explaining its physical effects on us, but we are all still wondering what causes it in the 40,000 year history of man.

In this chapter it is proposes such an operator based on the realization of a background of constant time explained earlier. It is essential to realize that we exist in 4 spatial dimensions expressed as [Ut, x, y, z] not the existing and historic 4D including [temporal time + x, y, z].

The ratio of 0's and 1's are for representation purposes only

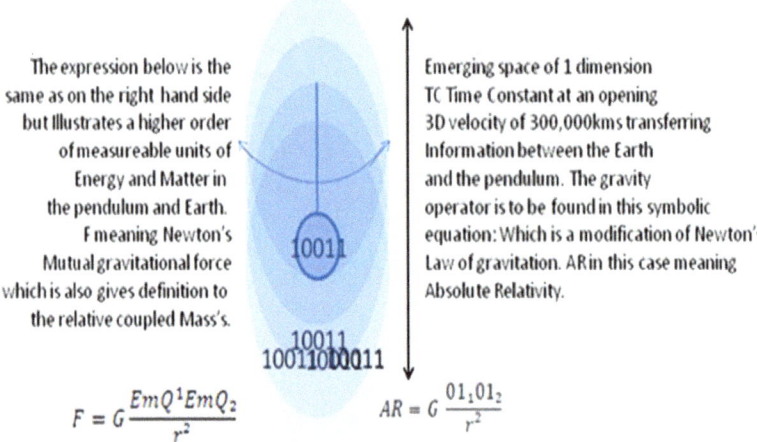

The expression below is the same as on the right hand side but Illustrates a higher order of measureable units of Energy and Matter in the pendulum and Earth. F meaning Newton's Mutual gravitational force which is also gives definition to the relative coupled Mass's.

Emerging space of 1 dimension TC Time Constant at an opening 3D velocity of 300,000kms transferring Information between the Earth and the pendulum. The gravity operator is to be found in this symbolic equation: Which is a modification of Newton's Law of gravitation. AR in this case meaning Absolute Relativity.

$$F = G\frac{EmQ^1 EmQ_2}{r^2} \qquad AR = G\frac{01_1 01_2}{r^2}$$

The above illustration shows an information couple between the Earth and pendulum. The repetitive swinging is a result of its conserved kinetic energy and relative mass caused by the information mutually conveyed from Earth to it and vice versa.

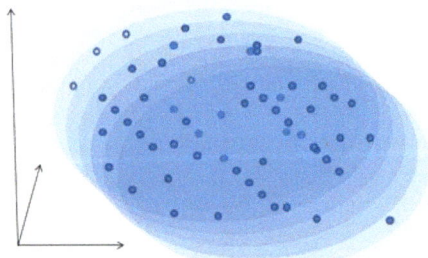

Manifold of Primary Space with an isometric construct of 1 dimensional Time points

Sir Isaac Newton's notion of the Absolute Background of Space extended to form the authors concept of its construction and emergence

In the above image one has to imagine it is entirely constructed in the manner earlier described and more like an expansive open celled foam comprising of regular 1D cells. The 3 arrows only indicate the 3D directions but the entire body exists as a framework of constant time and a uniform 1 dimensional plane. The manifold is fully populated with [0] time points which are represented by the sphericles but unlike the image are fully contiguous and homogeneous. New Space is constantly being emerging illustrated here by the additional oval 1D layers sets. The entire manifold continues to emerge endlessly and the entire system continues to enlarge in the virtual apparent 3 directional arrows. forming a manifold 1D plane of space.

The above illustration is 100% empty of matter and all the time points continue to be referenced as zero and unaffected by matter. The distances represented by the arrows in the illustration may be considered as infinitely sub micron or similarly unimaginable billions of light years. The conditions shall remain homogenous and isometric with respect to the numerical 3D value of the arrows.

If we modify this manifold by introducing matter to form two separate bodies, being a quark, atom, planet or sun etc, the presence of this matter now interacts with the 1D space point emergence.

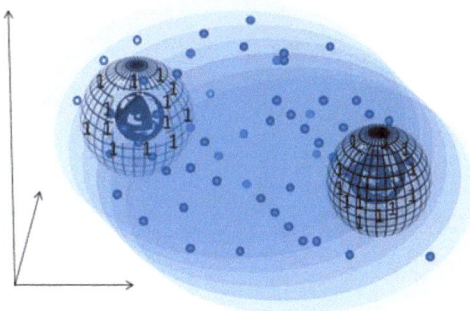

Now we have a new situation where we the emergence and population of the zero time points are being modified according to the local conditions within the bodies. Meaning each item of material no matter how small is creating a bits of information about the energy and number of materials each contain, which may have or not have mass related particle associated with it. A duplicate of each body is now under construction and conveyed spherically from it at the value of [C]. Not because the emergence of the 1D time points is governed by the speed of light but because this is the value at which New Space emerges in this part of the universe. A photon is only able to transmit through our universe because New Space is being constantly formed at this value.

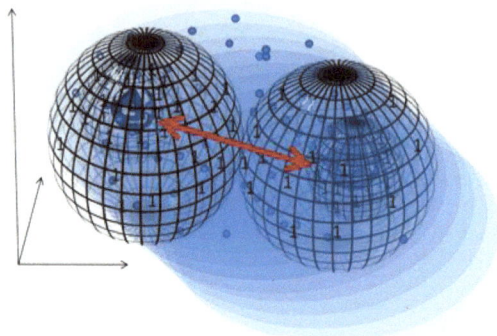

In this illustration is shows how the bits of information transit across 3D virtual spacetime and the initial conditions of the matter becomes contiguous with the other body. This can be an atom or planets billion of light years away. As soon as the informative spherical faces becomes contiguous then the attraction, the gravitation is initiated. Then depending upon the ratio of [ 0's or 1' ] of constant time information they contain will determine which has the dominant influence over the other. Hence have a greater or lesser gravitational affect according to Newton's classical modified law above. In the case of his apple the illustration below shows the same information dynamic.

Newton's apple is obliged to move in the direction of the Earth because there is no other greater information acting on it. The apparent weight of the apple is caused by the mutual pull of both sets of spherical information which is being updated by the second at the velocity of light through the medium of the Primary Absolute Space. When the apple was nothing more than a flower it had a reduced set of information about itself. As it grows on the parent tree it is swelling with the addition of more matter [$mQ$]. It is being purged with matter and updates information as New Space is being created though it. This causes the ongoing increase of gravitation between it and the Earth. The nformation of all apples that ever existed is persistent and remains as a package of information in the Primary 1D Space, as does all matter forms which it shares in 3D spacetime.

Historic statements from

Sir Isaac Newton:

*Principia*: "Absolute, true and mathematical time, of itself and from its own nature, flows equably without relation to anything external absolute space, in its own nature, without relation to anything external, remains always similar and immoveable."

Vs

Albert Einstein:

Matter bends space ( Special theory of relativity ) , and then tells matter what to do ( General theory of relativity ), there is no Newtonian absolute space, if there is, it is only of hypothetical interest.

If this theory is correct it must then be joined to Newton's. What he did not realize in the 17C is the fact that his Absolute Space is a dimension in its own right and having a construct of Zero Time and constantly emerging. Then our observation of Einstein bodies in 3D hence 'spacetime' is a separate entity and dynamic in temporal time, both are valid but separate. When considering events in 3D of spacetime one is obliged to use the measuring limits of temporal clock time. When one does it has to allow for the constantly updating information between the bodies.

More recently in the case of NASA Gravity Probe B its orbital distance from the Earth meant that it was receiving historic information from Earth, hence it appeared to twist space. On the basis that the information is transmitted at value [c] from any body. Despite the relative short distance of it from the Earth New Space has to emerge transferring that information to the Satellite from Earth.

This may only take a minute fraction of a second, however it cannot be avoided. There is a delay of information between the two. So the Gravity Probe B was responding to historic information about the Earth and not a simultaneous measurement.

## So how does Gravity information from one body attract another?

To cause 'gravity' between objects it is essential that the objects are informatively contiguous in 1D. Even though they may be separated by countless billions or even countless trillions or light years. Newton's apple was only a few meters from Earth, and it did not have to wait long before the exchange of information occurred and it fell to earth in front of him. He then wondered about that and wrote his law of gravity centered around the mass of the apple ( it's weight here on earth ). Then Einstein thought about that and extended it with Dr Mach and Professor Minkowski to create the concept of spacetime with a primary dimension of temporal time. His notions were centered around the Earth and 3D, Newton's was centered around the universe in 1D but did not realise it.

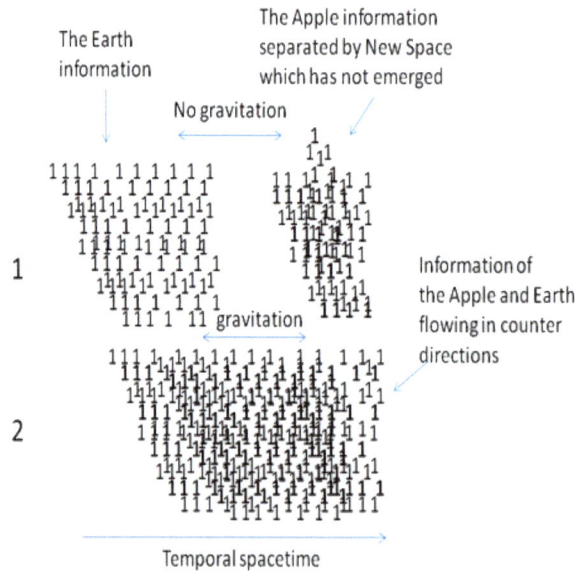

In condition 1 above New Space has not been created between the apple and Earth, in this case the apple is far distant from Earth. Both sets of information are in transit between them at the value of [c]. There is no opportunity for mutual gravitation as their information is separated by space which has not emerged. In condition 2 New Space has emerged and the historic information regarding both are now permanently entangled and updated in our experience of temporal time. In this hypothetical condition the apple may be light years away from Earth and as far as the Earth is concerned it is still swelling. But the apple with respect to its own temporal time has long since fallen and rotted. But the Earth is still receiving information that it is still hanging on the tree and continues to attract it gravitationally. It is reacting to historic information. The apple long since ceases to exist and produce any gravitation. They exist in unique envelopes of temporal time caused by their separation in 3D spacetime.

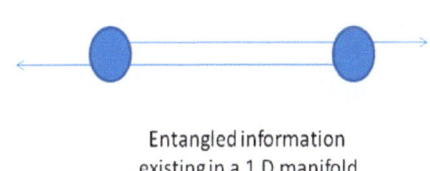

Entangled information
existing in a 1 D manifold

The outcome of mutual information entanglement is affiliation and influence of each respective body, but not to be imagined in 3D, only in 1D of the Primary Absolute Space. The information from each body continues to extend through and beyond the bodies illustrated by the arrows above. The outcome is a common gravitational chord binding the two bodies. In the illustration below. Newton referred to this as a rope tensor but could not explain it in a 1D coupled plane.

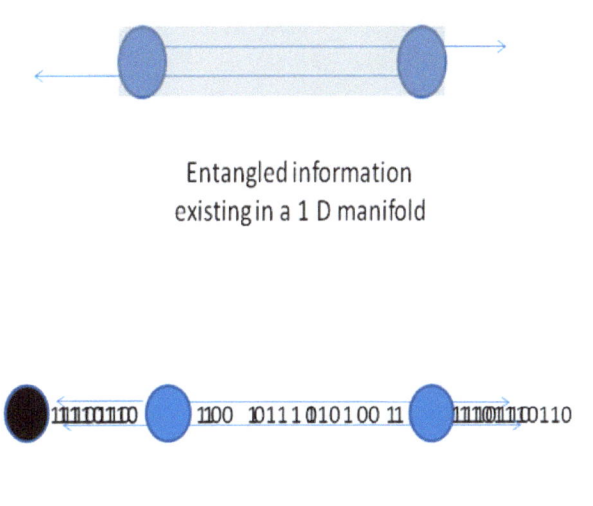

Entangled information
existing in a 1 D manifold

Information is exchanged and emerges beyond
each respective body and becomes additive

The net result becomes an amalgam of matter

Once the amalgam has formed; a new particle, planet, sun or black hole it has its own unique conditions, information and gravitational characteristics. The illustration below shows a potential ratio relationship of 0's and 1's created by its respective parent body. A small low energy body will have fewer 1's and greater number of 0's. Conversely a larger body, which may be more energetic such as a sun will produce a greater number of 1's hence the mathematical greater ratio and subsequent strength of the relative gravitational force.

The stream of information has a finite length in the plane of the blue arrows which connect directly with the parent body creating it. Because of the inverse square affect, a body which is strongly gravitational is diluted by the scale of distance, on the basis that the quantum of 1's are distributed over a wider spherical envelope when we envisage occurrences in spacetime of 3D.

The 0's and 1's are instantly affiliated hence the outcome of gravity produced shown below.

$$= AR = G \frac{01_1 01_2}{r^2}$$

The 0's and 1's are copies of the parent body signature and then represented in the larger order by measure of Energy and matter.

Where one body condition has caused the greater conversion of 0's to 1's it has the greater gravitational attraction than its partner body which will have a lower energy and matter content. Each and every sub quantum particle no matter how small will affect the change of adjacent 0 into a 1. Illustrated by the theoretical atomic layout below. The net effect is that the entire content of an atom and the huge space across it is identically mapped, and this information is constantly updated and conveyed into the Primary Dimension Ut. The huge 1D void existing between these bits of information remain as a 0's.

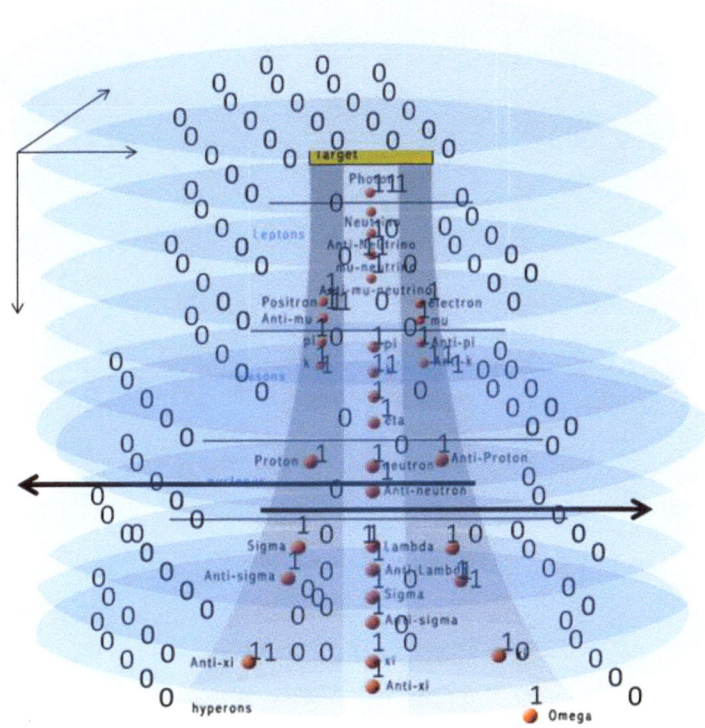

Absolute Space is emerging around every sub infinitesimal entity of matter in 1D at the value 'C'. The manifold enlargement is manifested as a 3D spacetime gravitation between bodies.

## 8 General Statements

1 The universe exists as 2 separate entities of time a. The Primary dimension[ Ut] which is constantly emerging and populating itself with zero 1D time points.

2 [Ut ] is a manifold existing with isometric time points of zero causing a 1 dimensional isometric field.

3 As [Ut ] is constantly producing New Space, if a body of matter exists in its vicinity, the particles of that body modify the character of the zero time points and become a [1] point of zero time.

4 New Space is constantly emerging which produces a constant stream of [0 or 1 ] where matter is apparent.

5 As New Space propagates itself information of each body and particle in it spacetime is distributed spherically which we refer to as 3D spacetime.

6 The expression for the Universe is Ar=Ut, x, y, z =0 or 1, or Ut3d=0 or 1

7 The expression for the operators [0's and 1 time points [0 → 1], [1 → 0] counter flow and oppositely handed .

8 The resultant gravity is determined by the ratio: [0 ∞ 1] in a contiguous body and the resultant gravitational force between 2 bodies:

$$AR = G \frac{01_1 01_2}{r^2}$$

The body which has the greater number of [1's] has the resultant larger gravity.

11 This symbolized expression refers to Absolute Relativity and route to 'The Grand Unification' Theory:

$$AR = Ut3D$$

# Gravity Explained Newton & Johnson vs Einstein & Minkowski

Newton's practical 'Absolute Space' philosophy vs Einstein's mathematical thought models of Special Theory of Relativity and General theory of Relativity

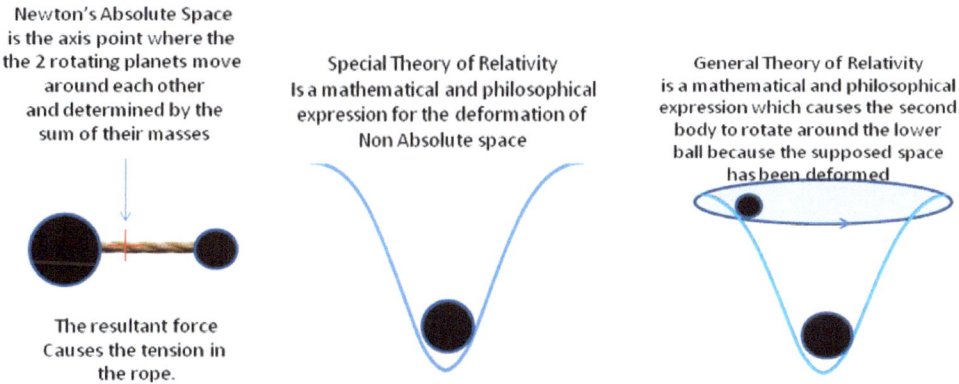

The strength of Johnson's notion is that Absolute Space exists as a plane of constant 1D time irespective to the motion of the bodies. Einstein's case is that the Absolute does not exist and space can be defined as spacetime which combines the ideas of Euclidian 3D + Minkosksi temporal time, and his mathematical philosophy does not require an absolute space for his to work. Since then we have discounted Newton, and continue to fail to understand the operator mechanism for gravitation.

What Newton did not realise is or could not realise at that the birth of his ideas of Absolute Space is being created second for second at the speed of light. As a totally independent singular dimension. Forming a forever enlarging manifold with a contruct of 1D points of constant time being produced ismetrically everywhere in the universe. And where matter exists causes the modification of these points, which determines the information corresponging to the bodies. Hence the gravitation between the bodies in his illustration above it is not a rope but information. Which hence causes the gravitation by mutual affiliation.

STR and GTR ingeniosly predicted but unable to provide an answer to the below obervations as it does not idenfiy a Gravitational Operator as a reference, it is merely a mathermatical model which can identify the paradoxes without presenting a satisfactory solution.

1. GTR & STR predicts but fails to explain:

    a. Temporal time dilation at high % Velocity of [C]
    b. Relative length contraction at high % Velocity of [C]
    c. Temporal time increase at increasing distance from a source of gravity
    d. Resolves gravity with the 3 other forces of Nature.

2. These are simply explained by this theory:

    a. Temporal time shortens the closer a clock approaches 'In Phase' velocity of [Ut] emergence. Currently 300,000kms = Time Constant [0] of the Primary dimension.

### Approaching speed of light in temporal time

### Emerging Absolute Space at velocity of light with a 1D manifold of Constant Time [0]

### The clock has stopped because it is in velocity phase with the emerging Absolute Space where time is Zero

b. Apparent length shortening occurs as the relative amount of emerged space is reducing with increasing observed velocity and less available space has emerged causing it to be relatively and apparently shorter. The information period is shortened not the train.

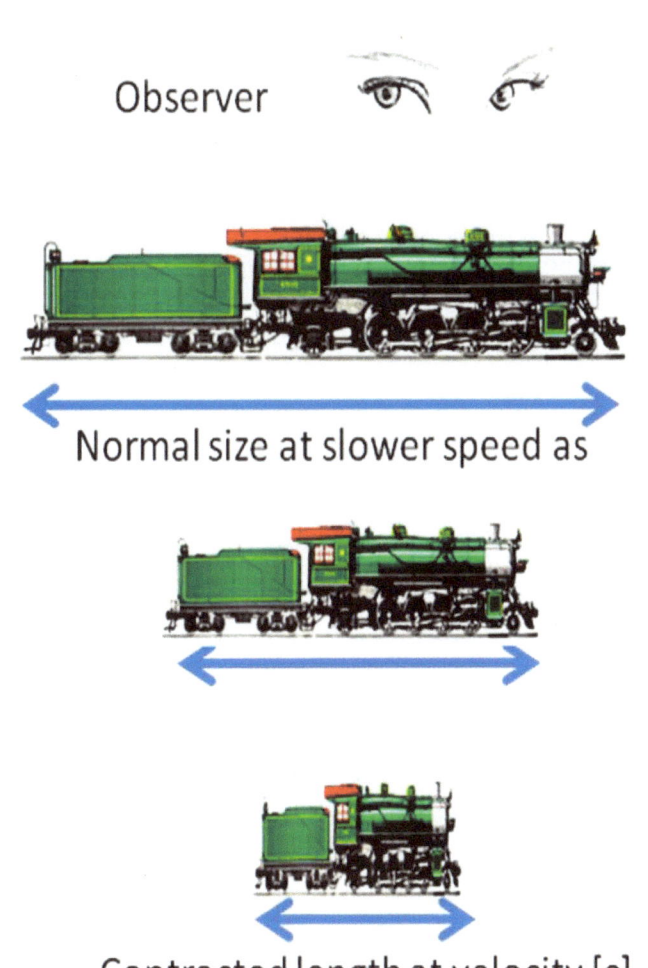

c. Temporal time increases moving away from a gravitational point and its field. When the clock on the surface reads 1 o'clock. The temporal time has an originating reference frame of information, and when conveyed beyond that origin e.g. the clocks at the rim, New Space emerging also has to convey the original information spherically outward in our 3D but has a construct of 1D.

However, as the emergence of new space moves the original information away from the surface it has to swell in according to the inverse square law as we witness in 3D. Illustrated by the spherical grid below. This grid is an exact solid duplicate of the earth. The Earth has 2 existences. 1. The real one which we enjoy, and 2. The virtual one which we are unaware of existing in 1D. The real one remains with a constant diameter composed of matter and energy, the virtual form has a diameter determined by the velocity of light since the very beginning the Earth was created.

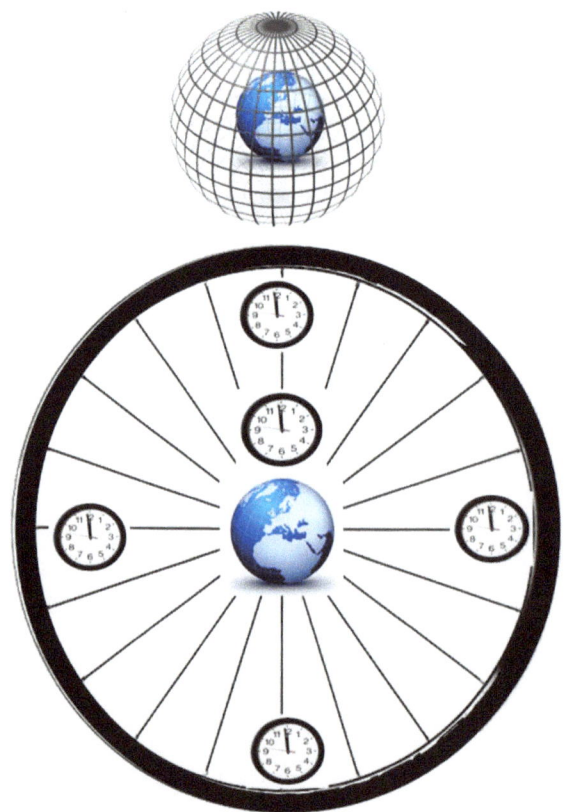

The clock has to tick faster and the a apparent time passes faster, because it is moving through an enlarging inverse distance with **diminishing gravitational information**. It is acting like the bicycle wheel. The further away from the axis the faster it has to tick. Conversely if you return the clock toward its origin the clock will show the opposite effect. The clock has its own unique field of information just like the Earth. The ratio of exchanged information when of the surface determines it has a local surface time. Then as it moves away the ratio of information changes in favour of the clock. Then the clocks local time increases and referenced to the diminishing affect of gravitation. Temporal time is determined by gravitational information.

Let me explain this another way so you can visualize what is happening it is very simple. Please refer to page 23 you will see Sir Isaac Newton's apple falling toward the Earth. The reason it is doing that is because it is receiving a more dense amount of information [ 0's and 1's from the Earth] This is what determines its local weight [relative mass] being so close to a major gravitational source.

If you put the same apple into shuttle craft as it moves away from the Earth it is receiving the same information but substantially diluted with increasing distance. The apple in isolation has the same information about itself as it did on the surface. The clock is exactly the same and will apparently tick faster as there is less information interacting and influencing it from its point of origin.

Should the shuttle craft then fly close to Jupiter the clock will tick at a different rate. It will slow down according to the ratio of information it is receiving from that planet. Temporal time passes at a different measured rates depending upon which planet you visit and directly related to the ratio of combined information of the respective bodies.

The temporal time on each planet will have its own unique local time. This is another reason why temporal time causes so much confusion, and the sooner it is disposed of as dimension the evolution of science can make leaps and jumps of new knowledge.

A further example if a spaceman from Jupiter a planet with substantially more gravity than Earth, where to arrive and land here, he would be very curious about his watch. Although it will show time passing, and observes the sun rises and sets more quickly, his watch will indicate **local Earth temporal time**. If he then phones his control station on Jupiter they will report that their seconds are ticking slower. Without the understanding of this theory he would be completely confused. That is why this phenomena causes so much confusion in the global scientific community. And reason why the Google navigational satellites have to constantly correct their time to that of Earth.

Temporal time is a slave to location and motion, as it is obliged to move through the manifold of 1D and interact with the emergence of New Space of constant time - the Absolute Space of [Ut]

It is not so difficult to understand these rules but we must realise that we exist in 4 spatial dimensions, the only dimensions and set up correctly in the first place. And quickly remove temporal time as dimension which has caused so much confusion since the beginning of the 20th century.

# 10 Conclusion

## The Primary operating mechanism

- Ut, x, y, z are the only 4 relevant operating spatial dimensions of the universe.

- Gravitation is caused by the exchange of information at the sub infinitesimal scale by modification of constant time [0] bits to form[ 1] bits by the constant emergence of New Space [Ut]. The resultant ratio of information determines which body has the greater or lesser gravitational force.

## The Secondary mechanism

- Constant Time is the prime dimension of the universe, the master and dictates everything occurring in the 3D.

- The kinetics involved with its emergence determine the causation of matter, it's removal before and after the Big Bang.

- While the value of its emergence is and around the value [C], light is obliged to be limited to this velocity in 3D and matter will persist.

- Should the kinetics involved with the creation of [Ut] emergence vary this will adjust the atomic cohesion of the standard model of particles. Should that happen the currently persistent 3D matter will revert back into the kinetic form and disappear and return to Newton's Absolute Space. In keeping with his intuitive thinking in the 17C and discounted by Einstein in 1905.

- It combines the 3 other natural forces.

The End and to the mystery of gravity, velocity of light

'The law of Unification'

All rights reserved no part of this book or theory may be copied or duplicated in whatever form without permission from the copyright holder 30th January 2013. Should any explanation of this theory be used in any shape for form this book and author must be cited

Edward Johnson

# Notes

Notes

Edward Johnson

Notes

www.ingramcontent.com/pod-product-compliance
Lightning Source LLC
Chambersburg PA
CBHW051107180526
45172CB00002B/803